How To Use 80% Less Water and Live Respectably

Legal Notice: The writer and publisher of this publication expressly disclaim all liability for the use or interpretation by others of information contained in this publication and/or listed Web sites. The author, publisher and distributors of this publication hereby disclaim any and all liability for any loss or damage caused by errors or omissions, (should these exist) whether such errors or omissions resulted from negligence, accident, or other causes. In this publication there may be inadvertent inaccuracies including technical inaccuracies, typographical inaccuracies and other possible inaccuracies. If legal advice or other expert assistance is required, the services of a competent professional person in a consultation capacity should be sought. the information contained herein may be subject to varying state and/or local laws or regulations. All users are advised to retain competent counsel to determine what state and/or local laws or regulations may apply to the user's particular business. The Purchaser or Reader of this publication assumes responsibility for the use of these materials and information. Adherence to all applicable laws and regulations, federal, state, and local, governing professional licensing, business practices, advertising, and all other aspects of doing business in the United States or any other jurisdiction is the sole responsibility of the Purchaser or Reader. The Author and Publisher assume no responsibility or liability whatsoever on the behalf of any Purchaser or Reader of these materials. Any perceived slights of specific people or organizations are unintentional. Products, services and websites' content vary with time. Please verify any published information.

© 2015 Phil Gurian

Hardcopy ISBN-13: 978-1514628263
Hardcopy ISBN-10: 1514628260

This book is comprised of three books that are normally sold separately. Your books are presented in this order:

1) Absolutely Essential Tips For Buying On eBay
2) How To Use 80% Less Water and Live Respectably
3) The Ultimate Collection of Resurrections and Rebirths

Table of Contents

Book 1

Absolutely Essential Tips For Buying On eBay

1) Last Minute Bidding Frenzies - Perhaps you've noticed that often there's a bidding frenzy in the last one minute of bidding. New bidders may suddenly start bidding in the hope that the previous bidders will not be watching or can't increase their bid in time. Often however it's because of *Sniping*.

Sniping websites automatically bid on your behalf, often in the last 10ish seconds. Simply sign up, enter an eBay item number and the maximum price you're willing to pay. Hidbid.com and goofbid.com offer sniping services that place bids for you.

Typically you'll need to give sniping sites your eBay password for them to work (ugh!!) Obviously that is a serious security concern.

There's little protection from eBay if things go wrong when sniping, since you willingly gave your password to a third party. If you do sign up for such a service, never use the same password for eBay as you use for other accounts like banks accounts or email addresses.

2) Second-chance Auction Scams, Beware of Them - Unscrupulous people sometimes watch bidders in high-dollar auctions and try to take unsuspecting buyer's money after an auction ends.

The scheme, known as a *Second-chance Auction Scam*, is just one of many types of Internet auction frauds reported to the *Internet Crime Complaint Center,* or *IC3*.

Second-chance scammers wait until auctions end and then offer bidders that lost, a phony second chance to purchase the item -- usually through a wire transfer service. This happens more often than people realize, beware!

3) Misspelling Search Tool - Typojoe.com, goofbid.com, bargainchecker.com, fatfingers.co.uk and baycrazy.com - There are many items listed on eBay every day that have misspelled words in the title. It's unfortunate for the seller but chances are good those listings will not come up well in eBay's search engine (because misspelling causes keyword problems) and thus not bring the seller top dollar. Their loss can be your gain!

4) Bidding Tip - Often sellers start auctions at .99 cents, (or at least under a dollar) hoping a bidding war will erupt. Many items go unspotted, staying at this super-low price (99 cents). *LastminuteAuction.com* hunts for eBay auctions due to finish within an hour but where the price still is very low.

With these items in particular, double-check delivery charges, as some sellers hope to recoup costs by charging a little extra (though eBay's now set maximum delivery charges for many categories).

5) Don't Forget About Facebook - Facebook Marketplace is a force to be reckoned with. Also sellers often are open to haggling. Just log on to your account at Facebook and search for "Marketplace". It's also worth checking to see if there's any local Facebook selling groups in your area.

6) Nigerian Type Scam for Paying. These unscrupulous people want to pay with a money order that they claim to already have handy. Often it's for more than the purchase amount. He writes to ask if the seller would be "honest enough" (or something of that nature) to send him the extra cash along with the item. (However he might just try to only buy the item with it and not ask for extra cash.) Unfortunately the money order can look okay but is counterfeit. They particularly like the *Buy It Now* feature.

7) Set Long-term Alerts For Rare Items - If you want something very specific or hard to find, set a 'favorite search' and eBay will email each time a seller lists your desired item.

Simply type a product in eBay's search bar, such as "silver dollar", and click 'save search'. Be as specific as possible for the most accurate results. When (and if) someone lists one, you're alerted with an email.

8) Don't Assume eBay's the Cheapest Place To Get Your Item - Many people assume that if it's on eBay, it's automatically the least expensive place to get it, but that often isn't the case. Perhaps you'd also like to use *shopbots* (shopping robots) that check numerous Internet retailers to find the best price. Type into a search engine "shopping comparison sites".

The same rule applies when buying used merchandise. Check used marketplaces on Amazon.com and Play.com - you may even get it for free on *Freecycle* or *Freegle*.

9) Check the eBay Going Rate For an Item - There's a quick way to check an eBay product's average price. Enter the item into the search box and click "completed listings". What will come up is a list of prices that similar auctions have already settled on. After that, sort it by "Price: lowest first". If the price is red, it means no one bought it. Green means it sold. Figure out the average price.

10) eBay has banned the selling of intangible items, and that includes curses! - Among the items that were prohibited as of August 30, 2012, are "advice; spells; curses; hexing; conjuring; magic; prayers; blessing services; magic potions; healing sessions; work from home businesses and information; wholesale lists, and drop shop lists."

11) Haggling on eBay Can Pay Off - There's nothing wrong with asking for a discount, even if the listing doesn't have the "make offer" indication. Haggling works best on *Buy It Now* listings, or auctions with a high start price and no bids. Also you'll likely do better if you haggle as the auction is coming closer to closing as the seller could start feeling more desperate.

To contact the seller, click on the seller's nickname then "ask seller a question". If you're polite, you'll likely get further. Blunt requests such as "dude, how about $15?" likely won't work out as well. Remember the seller is likely going to lose money doing this so no point in being annoying.

Once you've arranged a deal, try to keep the transaction within eBay. Ask the seller to add (or change) a Buy It Now price. That way you don't lose the usual eBay buyer protection privileges.

12) Other Things to Do To Exploit Sellers' Screw-ups - Some sellers make basic mistakes, leavings goods going for bargain money.

As well as spelling boo-boos, another error is to leave out key details such as shoe size, dress brand, saying a console's an a Wii when the photo shows a Xbox. At this point, many buyers give it up as "too much hassle".

So contact the seller to fill in gaps, but don't ask the question via the item's listing page, (because that way, when the seller replies, eBay lets them add their reply to the main listing, so it's no longer your secret.)

Instead, ask the question via the seller's profile (make it clear which item you're talking about). They might not bother with the extra hassle of adding it to the listing, so you'll be the only one in the know.

Also the seller might not realize how pricy an item he/she actually has.

13) Tool to Track Down Crazy End Times - Listings that finish at anti-social times often get fewer bids, thus sell for less. To locate auctions that finish when fewer people are around to bid on them, use BayCrazy's *Crazy End Time* search. (A lot more on the best times to end your auction in the next section of the book *"Selling on eBay"*.)

Check out their auto-bidding tools if you don't want to spend all that time in front of the computer bidding at odd times. Other BayCrazy.com tools include "unwanted gift" and "ending now" searches. http://www.baycrazy.com/nightowl (Baycrazy offers other eBay related opportunities also.)

14) Search Descriptions as Well as Titles - eBay automatically searches seller's titles for results that include your specified keywords. If you're not getting the results you want, try also searching the item's *description* too. (To do this go to Advanced Search.)

For example, imagine you were searching for a REI Jacket. Unfortunately the seller may be selling one but only put "Ski Jacket" in the title however he mentioned "REI" in the description. Include description in your search and then it should then come up.

15) Search Using eBay Boolean Logic - If a seller could describe an item different ways, you can make eBay search for several different ways of describing it at once. Just place "((" at the beginning and enter different phrases individually enclosed by quotation marks, then followed by commas.

So for example, type... (("fishing tackle", "hook", "reel" ...and it will simultaneously bring up listings that contain the words "fishing tackle", "hook" and/or "reel".

16) Add A Few Extra Cents to Your Bid - When bidding, you enter a "maximum bid", and eBay makes automatic bids on your behalf up to your maximum bid.

Don't enter a round number. For example, if a coat is currently selling for $20, and the most you are willing to pay for it is $25, enter a maximum bid of $25.24. If someone tries to outbid you by entering the round number of $25, they will receive an outbid notice. eBay will go your bid, even though it's just 24 cents more.

17) Be Somewhat Skeptical of Feedback - eBay sellers have a feedback rating that acts as a useful guide to previous seller's opinion's of them. As a guideline, look for a seller with more than 98% positive feedback and a high feedback score of at least 30. Also ensure you read their feedback from their *selling*, not just their *buying*. (To see their feedback, click on their username).

18) Seller with Zero Feedback Could be Cause For Concern - Think twice before purchasing expensive items from a seller with zero feedback.

Remember feedback's useful but not infallible. One thing to watch for is traders selling a number of cheap things for $1ish each to build their feedback, and suddenly listing items costing hundreds each.

19) Check to Make Sure You're Bidding on the Actual Item - Sometimes you assume you're bidding for an item on eBay (or any auction site,) when all that's actually being sold is a link to another site selling it. People are not suppose to be able to sell these on eBay but they can fall through the cracks.

Always read the whole description in detail before bidding. Often the catch is hidden in the text at the end – an attempt to protect the seller from any recourse.

20) Scam - Beware of it - It's a red flag if a seller writes "Before bidding, contact me" then asks for a money transfer. Thieves who hijack actual eBay accounts might use this tactic.

21) Scam - Beware of it - Always be worried if you're asked to pay by an instant money transfer service such as Western Union or MoneyGram. Instant money transfer payments cannot be traced and are highly popular with thieves.

22) Sneakily Find Underpriced Buy It Nows - Feel free to hunt for Buy It Now bargains also. Perhaps the seller under-values their item making their price a good deal.

These steals are snapped up quickly. Go to "Advanced Search", select a category you're interested in, filter it to show *Buy It Now* items and sort the results.

23) Complain ASAP – You may want to open a case if you're unhappy with your purchase. (There are some exceptions such as tickets for events that are months away.) Read more on eBay's protection policy.

http://pages.ebay.com/help/policies/money-back-guarantee.html

Under eBay's own Buyer Protection rules, buyers are eligible for a refund if the item's "not as described", meaning it didn't match the seller's description. http://pages.ebay.com/ebay-money-back-guarantee

24) Pay by PayPal - Avoid sending checks and never use money orders. It's much harder for scammers to disappear with your cash when you use eBay's online payment system, PayPal.

Paying this way costs the same as paying by check, but means you're covered by eBay's Buyer Protection program. If an item is faulty, counterfeit or non-existent, you are far more likely to get a refund.

25) Outbid? Don't Give Up On It Yet - Missed out on a desired item by pennies? Don't give up hope. As every seller knows, sales sometimes don't materialize when buyers change their minds or can't come up with the dough. Because of that feel free to send a friendly message such as: "Hi, I've been looking for this poster for years and just saw your finished auction. Please let me know if the sale doesn't come through."

They may send a *Second-chance Offer*, which are sent out by sellers to unsuccessful bidders if the winner fails to pay up. Ask them to relist at an agreed *Buy It Now* price.

26) Know Your Consumer Rights - When buying from a person who makes or sells goods for resale on eBay you often have the same rights as when buying in person from a shop that does the same. This means your goods must be of satisfactory quality and as described.

With private sellers it's buyer beware. Buyers' only rights under law are that the product is fairly described and the owner has the right to sell it.

Under eBay's own Buyer Protection rules, buyers are eligible for a refund if the item is "not as described", meaning it doesn't match the seller's description.

27) Beware of All The Fakes - While eBay has a 'flag and remove' policy to help identify fakes, still plenty fall through the cracks.

If you're buying big-name brands, do your research first. Carefully check sellers' feedback and post on the forum's eBay board to get others' opinions.

Be especially wary of overseas sellers or branded items that seem especially cheap.

The more *unprofessional* the photos, perhaps the better. Thieves often take professional photos from the brands' sites. Legitimate sellers typically take photos of items at home that might not come out as well.

28) Think Twice Before You Give A Seller Negative Feedback - Of course, negative feedback is often justified but have a heart, don't leave negative or even *neutral* feedback without first trying to work the issue out with the seller. Most sellers are good folks who will try to help particularly, as it can mean a lot to their business to stay in your good graces.

Remember eBay users can view the feedback you've left for others, and if you leave a significant amount of negative feedback, they may well decide you're too high of a risk to sell to.

29) Add An Item You're Interested in to eBay's "Watch List" - Want to keep track of an item without bidding on it? eBay lets you add items to a "Watch List", so you can relax knowing you'll get an email reminder within 36 hours of the auction ending. To watch an item, just click the *"add to watch list"* link in the upper part of the item's eBay webpage.

30) Don't Do Private Purchasing - Sellers may suggest you do a deal outside eBay for a cheaper price. If you do you'll likely have less protection if things go bad. You won't be able to leave negative feedback and you won't be protected by eBay's Buyer Protection Plan. At the least buy it using Paypal for the buyer protection.

31) Think Safety When Picking Up An Item In Person - The usual precautions apply. If you get to their door and the seller's holding a butcher knife, now's the time to run.

32) Think International - There's bargains to be had on overseas eBay sites. To include foreign auctions in search results, click "worldwide" for location.

Still can't find what you want? Another option is buying directly from *international* eBay sites. The main ones are USA, Canada, Australia, Germany, France and Spain - there's a full list at the bottom of eBay's homepage. Make sure that the item reads "Shipping to: *worldwide*" before bidding as some international sellers only do business with their country's buyers.

Always factor in postage and if applicable, custom fees. Remember that return postage fees could be hefty. Also what kind of credit card protections will there be? You're often still protected by eBay and PayPal's buyer protections (if you use PayPal), but it's worth investigating. Type in "buyer protection" in PayPal.

33) Don't Forget The Online Classified Ads - Again, let's not assume that because it's on eBay, that's where you'll get the best price for an item. Unfortunately that's often just not the case. Type "top classified ad sites" or something of that nature, into search engines. There's also *Freecycle* and *Freegle*. (Those two sites offer free stuff. freecycle.org and ilovefreegle.org.)

Remember, anyone can post on these classified ad sites. If someone asks you to pay by MoneyGram or Western Union, as always be concerned. It's a bad way to pay.

34) Check Other Auction Sites Also - There are other auction sites that can be found through search engines. If you're searching for something specific, it's also worth adding it to your search. *Auctionlotwatch*.com is a useful shopbot for online auctions. Search for an item and it trawls the big auction sites for you.

35) Join eBay Forums - Ask questions about anything, selling, buying etc. Great information is posted already and could be of use. Work together as a team. Find eBay and other auction forums by looking those up in search engines. eBay has forums also. http://community.ebay.com

36) eBay Research Tool 1 - To help in your research about selling items, you can go Type into a search engine "best selling eBay items." eBay provides that information.

37) eBay Research Tool 2 - You can use Ebuyers (www.ebuyersedge.com) to just search eBay for items as well as set up a saved eBay search (or a number of them). You'll get alerted with an e-mail when a matching item is listed.

38) Check Cashback and Voucher Websites - Check cashback websites to see if there's money back available on your eBay purchase. Type into search engines: "cashback and voucher sites".

Cashback sites give you a cut of their proceeds by setting you up with product and/or service providers.

39) eBay has trained teachers that could be in your area. Also see eBay University. Check out:

http://pages.ebay.com/sellerinformation/learn-to-sell-online/ebay-university.html

The End

Book #2
How To Use 80% Less Water and Live Respectably

Legal Notice: In this publication there may be inadvertent inaccuracies including technical inaccuracies, typographical inaccuracies and other possible inaccuracies. The writer and publisher of this publication expressly disclaim all liability for the use or interpretation by others of information contained in this publication and/or listed Web sites. The author, publisher and distributors of this publication hereby disclaim any and all liability for any loss or damage caused by errors or omissions, (should these exist) whether such errors or omissions resulted from negligence, accident, or other causes. If legal advice or other expert assistance is required, the services of a competent professional person in a consultation capacity should be sought. the information contained herein may be subject to varying state and/or local laws or regulations. All users are advised to retain competent counsel to determine what state and/or local laws or regulations may apply to the user's particular business. The Purchaser or Reader of this publication assumes responsibility for the use of these materials and information. Adherence to all applicable laws and regulations, federal, state, and local, governing professional licensing, business practices, advertising, and all other aspects of doing business in the United States or any other jurisdiction is the sole responsibility of the Purchaser or Reader. The Author and Publisher assume no responsibility or liability whatsoever on the behalf of any Purchaser or Reader of these materials. Any perceived slights of specific people or organizations are unintentional. Products, services and websites' content vary with time. Please verify any published information.

How To Use 80% Less Water and Live Respectably

Table of Contents

The hyperlinks in this book are listed online for easier access.
http://www.itvinstitute.com/wc-links.html

How To Use 80% Less Water and Live Respectably

The hyperlinks in this book are listed online for easier access.
http://www.itvinstitute.com/wc-links.html

A. Introduction. (Unfortunately it's a lot more than most of us think!)

According to a March 3, 2014 article in the Los Angeles Times, Americans *use twice as much water* as we think.

Though it may seem hard to believe, every day, each person in a family (for our purposes a family of four,) uses an average of 80-100 gallons of water. (This is as of October, 2014. Reference - The U.S. Department of the Interior/U.S. Geological Survey. http://water.usgs.gov/edu/qa-home-percapita.html.)

Including the water I re-use (recycle,) my average usage is much, much less, roughly **11 gallons** a day. (Grey water use, i.e. re-using rain water, is *not* discussed in this book, but a very worthy cause!)

In this book I often reference an extensive 2011 report by The California Department of Water Resources (along with the Irvine Ranch Water District [IRWD]). It postulates that daily the average California household used (at least at the time of its writing) more than 360 gallons of water. (http://www.irwd.com/images/pdf/savewater/CaSingleFamilyWaterUseEffici encyStudyJune2011.pdf).[1]

According to that 2011 California report, indoor water use accounted for more than 170 gallons per household per day; bathing, the toilet and/or leaks typically being what used the most water.

That extensive report also notes that the average California home recorded over 57 '*faucet events*' per day (IRWD page 30.) A *faucet event* is when a person opens a faucet and water comes out of it. I estimate I average 15 or fewer 'faucet events' per day.

I look respectable and have my own home. At home I have running water, a microwave oven, air conditioner, refrigerator, internet, cell phone, satellite HDTV, a surround sound audio system, DVD player, VCR, stove, oven and a car.

B. Overall Average Daily Water Usage Statistics Per Person in a Family of Four

According to the official estimates noted (and in some cases educated estimations,) the following is the average amount of water used by Americans for the various activities noted. (Please note, as you know average water use is not the same for everybody. The following numbers are not meant to be definitive.)

The following boldface daily water usage amounts are **per person** in a family of four, two adults and two kids. This family is living together.

The hyperlinks in this book are listed online for easier access.
http://www.itvinstitute.com/wc-links.html

1. Toilet Flushing – 18 gallons a day per person (AWWA Research Foundation).

TIP - By putting sizable, solid rocks in the toilet water storage tank, (not those which could crumble,) or sand-filled capped jars, less water is used.

TIP - Don't flush just for a tissue, cigarette butt or other small bits of trash. It wastes a lot of water and is one of the reasons toilet water usage is so high.

More efficient toilets use less water but many older toilets even use 3.5 to 7 gallons a flush! (A WaterSense labeled toilet is more efficient. These use 1.28 gallons per flush or less. WaterSense is a partnership program sponsored by the U.S. Environmental Protection Agency.)

2. Hand Washing – 5 gallons a day per person.

(*This doesn't include hand washing during a shower/bath.*) – A bathroom faucet typically expels water at two gallons per minute. (http://www.epa.gov/WaterSense/pubs/indoor.html.) Each washing takes about **a gallon** according to the U.S. Department of the Interior; U.S. Geological Survey website - http://water.usgs.gov/edu/qa-home-percapita.html. (It's less if one turns the water off when rubbing their hands with soap.)

3. Face Washing – 1/3 gallon a day per person.

(*This doesn't include face washing during a shower/bath.*) – Each face washing takes about **a gallon** of water according to the U.S. Department of the Interior; U.S. Geological Survey - http://water.usgs.gov/edu/qa-home-percapita.html. It's estimated that people wash their face when not bathing once every three days (on average.)

4. Hair Washing – See *Showering/bathing*.

5. Hair Washing Using Bottled Water.

As their tap water is too mineral-encrusted, some folks use bottled water to wash and rinse their hair. This use is typically one to three gallons of bottled water a day, but as this type of water usage is not the norm, let's not include it in our overall tally.

6. Shaving – 1 gallon a day per person.

As the sink is often filled up with a significant amount of water, and additional water is used to clean the shaver, numerous estimates put it in the *3/4 to 3 gallon* range.

7. Leaving the Water Running When Cleaning Vegetables, Fruit, Peeling Onions etc. – 1/2 gallon a day (on average) per person.

This can use a surprising amount of water. To have the water better clean the items being washed, people often increase the volume and intensity of the water flow.

Vegetables and fruit typically have pesticides, bacteria and other contaminants on them. Washing and/or soaking them is important. Who knows if the picker/processor recently washed his/her hands.

TIP - Many use a vinegar/water solution.

8. Teeth Brushing – 1 gallon a day per person.

The U.S. Department of the Interior; U.S. Geological Survey estimates it's **one gallon**. http://water.usgs.gov/edu/qa-home-percapita.html

Folks who brush their teeth (likely a good idea) multiple times a day bring up the average.

According to an EPA site, turning off the water as you physically brush your teeth can save up to 8 gallons of water each brushing! http://www.epa.gov/WaterSense/kids/tap-off.html. (Other experts argue the number is closer to 1-2 gallons, depending on the flow rate of the faucet used.)

9. Showering/bathing – 18 gallons a day, per person, on average. (Page 29, IRWD). *(Note, that average includes days when no shower/bath is taken.)*

The average number of gallons per shower, per person, is just over **18 gallons** (2007 statistic).

The U.S. Department of the Interior; U.S. Geological Survey postulates that a 'full' bathtub uses roughly 36 gallons - http://water.usgs.gov/edu/qa-home-percapita.html) Then often additional water is used for rinsing off. Because of all this baths tend to use more water than showers.

The majority of a shower's water flow is typically at or below the 2.5 gallons per minute standard set in the 1992 EPAct.

10. Drinking Water/liquid, and Water/liquid Used in Cooking (*Not including water for making coffee, cleaning vegetables, fruits, and running warm water to defrost frozen food.*) – **1 gallon a day per person**. (It would be higher if everybody was cooking their own meals, but chances are one person is often cooking food for everybody in the family.)

It's recommended to drink a half gallon of liquid a day. This includes soda, coffee, tea, juices and the liquid in the food we eat. (There is debate on how much of non-water liquids can be used interchangeably with water.) Of course there are circumstances such as sweating a lot, which will make your body require more, or less, daily hydration.

We use water for cooking rice, oats, noodles etc. We also use it for making frozen blended drinks, and a lot more.

11. Laundry (Clothes Washing) – **5.5 gallons a day per person**.

Annually the average American family washes roughly 400 loads of laundry. (Alliance for Water Efficiency. http://www.home-water-works.org/indoor-use/clothes-washer). High efficiency washers, in particular front loading washers, use significantly less water, (even as low as 15 gallons a complete load,) but as of this writing most washers are of average efficiency and use 25- 30+ gallons of water a complete load. *So let's say the average American washer uses 20 gallons of water per complete load.*

Multiply 20 gallons per load by 400 (the average number of annual washing machine loads per family) and the total gallons of water used per family is 8000 gallons of water. Divide 8000 gallons by 365 days and that's *22 gallons of water used per day per family averaged over the year.* But as there are 4 people in the family we need to divide 22 by 4 which computes to 5.5 gallons used per person (on average) daily in the family.

12. Coffee Making and Coffee Pot Washing – **1/4 gallon a day per person.** (As a reminder, these statistics are based on a family of four, two adults and two kids. So the water used for coffee making and apparatus cleaning would be ¼ gallon times four, which is a gallon of water a day.)

Coffee is often made for and shared by all adult members of the four person family. If a half gallon is the amount of coffee made daily then the remaining half gallon is used daily to clean the coffee maker and pot.

13. Dishwashing – (Page 259-60, IRWD) – Annually the average American family's dishwasher washes roughly 110 loads of dishes. An older model dishwasher will use approximately 10 to 15 gallons of water per load. Newer ENERGY STAR® dishwashers will use 5.5 gallons or less per load. (2011 statistic, Alliance for Water Efficiency, http://www.home-water-works.org.)

Statistically the use of an automatic dishwasher *often reduces* home water usage, (even though manual pre-rinsing of dishes is often involved.) This is because:

(1) We often leave the water running continually for washing and rinsing purposes when manually washing dishes
(2) We tend to wash dishes manually more often *(such as after most meals)* as compared to when we fill up and finally run a dishwasher
(3) Putting a lot of water in the sink and soaking dishes in-between washes can happen more often if manually washing dishes.

Not pre-washing dishes prior to loading the dishwasher could save 10 gallons or so a load. A lot of pre-washing can even reverse things and make the automatic dishwasher the least water friendly of the two, particularly if it's an older dishwasher.

TIP – Scraping the dishes instead of pre-rinsing might work well depending on your automatic dishwasher.

TIP – Make sure the dishwasher's fully loaded for the most efficient water usage!

As of August 11, 2009, ENERGY STAR certified dishwashers were required to use only 5.8 gallons of water per cycle (or less.) Older dishwashers however use much more water. A dishwasher purchased before 1994 uses *10 more gallons of water in each cycle* than a new ENERGY STAR qualified model.

Energy Star states that using an ENERGY STAR qualified dishwasher instead of manually hand washing the dishes, would save you 5,000 gallons of water annually! http://energystar.supportportal.com However as of this writing Americans used significantly *less* ENERGY STAR certified dishwashers than the more inefficient older dishwashers.

A) *Automatic Dishwasher Water Usage* – **1 gallon a day, per person in the family**. Automatic dishwashers use (on average) 10 gallons a complete load. As you know a complete load includes washes and rinses.

If the dishwasher is operated 110 times a year, that's about *once every three days.* If each full wash and rinse cycle uses 10 gallons, that's about 21 gallons used in a week. Divide that by 7 days and that's an average of three gallons a day. But let's estimate that an additional 3 gallons of running tap water are used to pre-wash/rinse all the dishes per load. Weekly that's six more gallons. Add that to the 21gallons and that makes a total of 27 gallons a week used for the family automatic dishwasher. Divide that by 7 days in a week and we have just under four gallons a day used for automatic dishwashing…but assuming there are four in a family then we need to divide that number by four. *So per person, the daily estimated average amount of water used by the average automatic dishwasher (per person in the family) is 1 gallon.*

As you know families can run their dishwashers more than once every three days. That will use more water than the above figures, depending on the efficiency of the dishwasher.

B) *Water Usage When Manually Washing Dishes* – 8-27 gallons a day is used for an average of 17.5 gallons…but assuming there are four in a family then we need to divide that number by four. So per person, the daily amount of water used in manually dishwashing is roughly **4.5 gallons.** (http://water.usgs.gov/edu/qa-home-percapita.html)

The amount of water used when manually washing the same number of dishes/pans etc. as the previous mentioned automatic dishwasher, depends largely on the intensity of the faucet water flow, the frequency of your dishwashing, the speed you wash with, and if you leave the water on continuously.

TIP - Pre-soaking dishes first can save water and make cleaning easier.

14. Utility Sink – The average indoor faucet generally runs at around two gallons per minute. As this type of water usage is not the norm, let's not include it in our tally.

15. Thawing – Thaw something in the refrigerator overnight instead of running hot water over it. As this type of water usage is not the norm, let's not include it in our tally.

16. Garbage Disposal and/or Manually Cleaning Out the Kitchen and Bathroom Sinks/drains – **1/2 gallon a day per person.** Garbage disposals use approximately four gallons of water a minute. (http://www.swfwmd.state.fl.us/publications/files/daily_water_use.pdf)

Homes with garbage disposals actually use LESS water on average than homes without garbage disposals. This may be because there are more clogged drains in homes without garbage disposals and a significant amount of water is used to clear drains. (Page 260, IRWD)

17. Cleaning/clearing Out Other Household Drains – As this type of water usage is not the daily norm, let's not include it in our tally.

18. Carwash – **1/2 gallon a day per person.** A commercial car wash that doesn't recycle any of its water typically uses dramatically more water than a commercial coin-operated self-service car wash, but washing your car in your driveway typically uses more water than either of them!

Washing your car at home with the water running fairly hard uses 80 to 140 gallons of water every ten minutes! However washing a bike, car or boat by primarily using a bucket of soapy water potentially saves a lot of water! http://www.imagesautospa.com/Environmental.html.

The Maryland Department of the Environment reminds us that we use 100 gallons of water with only a 10-minute car wash, assuming the water continues to run while we scrub.

As you know business' tend to wash their cars more often due to heightened appearance concerns and those dealing with salt on the road likely will also.

A family often has multiple cars and washes at least one of their cars monthly.

Let's just say only one of the family cars is washed monthly. That's 80 gallons used during a car wash in your driveway (if the water isn't on too long or going too fast.) 80 gallons divided by the days in a month is *2.6 gallons a day*. But there are four our family so we need to divide that number by four, which gives us a bit over a half gallon a day per family member.

19. Hosing off the Garage, Driveway, Roof, House, Floors and/or Sidewalk – 1.25 gallons a day per person in the family. In 60 seconds, a typical five-eighths inch garden hose running at the normal household pressure of 50 psi (pressure per square inch) expels 10-16 gallons of water. (http://www.stormwater.cecs.ucf.edu) The Maryland Department of the Environment tells us that 10 gallons are expelled every minute. Another water conservation reference puts that number at 14 gallons. (http://water.usgs.gov/edu/qa-home-percapita.html) The University of Central Florida's Stormwater Adademy states that hosing off a driveway uses an average of 130 gallons for every 15 minutes.

Let's say one of the above listed areas (perhaps the driveway) are washed on average once a month. 10 gallons are expelled every minute (Maryland Department of the Environment). Let's say the hosing lasts 15 minutes, that's 150 gallons. Divide that by 30 days (a month) and that's 5 gallons a day…but there are four people in our family so we need to divide that number by four, and that's 1.25 gallons used per person per day (on average.)

20. Hosing Myself, Kids or Others Off With the Garden Hose – As noted earlier, in 60 seconds, a typical five-eighths inch garden hose running at the normal household pressure of 50 psi (pressure per square inch) expels 10-16 gallons of water. (http://www.stormwater.cecs.ucf.edu) As this type of water usage is not the daily norm, let's not include it in our tally.

21. Sprinklers and Home Irrigation Systems (including personal gardens) – **.5 gallons a day per person.** (Page 260-61, IRWD) – Outdoor water use is primarily for landscape irrigation (such as watering the lawn.) The referenced IRWD study assumed that the percentage of homes that practiced landscape irrigation was 87%. Over-irrigation is a huge problem and one of the main ways households waste water. The IRWD study found that a savings of nearly 15% of total single-family use could be achieved simply by cutting the number of over-irrigation offenders in half.

Another respected reference reminds us that outdoor watering uses an average of two gallons a minute. http://water.usgs.gov/edu/qa-home-percapita.html

As you know there is more outdoor watering in the warmer, drier months than the colder wetter months. Let's say the average length of outdoor watering is 20 minutes a day and done twice a week for 22 weeks. That's 22.5 waterings of 20 minutes each. That's 22.5 x 20 {minutes} x 2 {gallons per minute} equaling 900 gallons a year. Divide 900 by 365 = 2.5 gallons a day (when averaged over the entire year.) But there are four in our family so we need to divide that number by four. 2.5 divided by 4 is a little over a half gallon per person per day.

(Please note that many feel this number should be higher, but growing season lengths vary greatly in different areas of the country. In areas where there is drought this number likely is higher.)

22. Leaks – **7.75 gallons per day per person in the family of four.** (Page 29, IRWD) – The IRWD study found that roughly 18 percent of wasted water is lost due to home leaks; that's 31 gallons per household per day! According to the EPA, if your toilet leaks you can be wasting 200 gallons of water a day! (http://www.epa.gov/WaterSense/kids/fixleak.html)

"Leaks" are major and often continuous water wasting events. The worst offending homes dramatically raise the average for everybody.

Incidentally, it should be noted that 'leaks' in this category includes leaving the sprinkler, garden hose and irrigation systems on *too long*. It also includes home swimming pool leakage, broken valves or couplings, faucet drips and the toilet flapper sticking.

23. The Running of Tap Water to Cool the Water Off For Drinking – As this type of water usage is not the norm, let's not include it in our tally.

TIP - Keep water in the refrigerator instead.

24. Water for a Pet to Drink. Also Water Used to Clean the pet and its Messes – 1/4 gallon a day per person in the family.

Daily, a healthy dog drinks between ½ and 1 ounce of water per pound of body weight. So a healthy 65-pound dog could be drinking between ¼ to ½ gallon of water daily. It slobbers and spills a bunch too.

Note, that gallon of water is for only one pet. As we are assuming there are four in our family then we need to divide that gallon by four, which makes a ¼ gallon per person.

25. Extra Water Usage Associated with hosting Parties and Other Social Events – **Undetermined.** I know this is taking the easy way out but I couldn't find data supporting a good enough estimate. It likely is a significant amount though as we use water preparing for and catering to guests.

26. Putting Water in the Car Radiator – As this type of water usage is not the daily norm, let's not include it in our tally.

27. Public Supply and Facilities – (Such community pools, water fountains, restaurant toilets and patron's complimentary drinking water) – As there could be so many residents and visitors in a locality who use the pool or water fountain, let's not include it in our tally.

28. Agriculture and Ranching – (Page 231, IRWD) – Livestock water use alone was an estimated 197 million gallons a day (2005 statistic). As this isn't a statistic for individuals, let's not include it in our tally.

29. Industrial Aquaculture – (Page 231, IRWD) – 646 million gallons a day (2005 statistic). As this isn't a statistic for individuals, let's not include it in our tally.

30. <u>Mining</u> – (Page 231, IRWD) – 53.1 million gallons a day (2005 statistic). As this isn't a statistic for individuals, let's not include it in our tally.

31. <u>General US Industry & Manufacturing</u> - "In the early 1900s, American industry used about 10 to 15 billion gallons of water a day. With the huge growth in industry following World War II, the industrial use of water also grew. By 1980, industry was using about 150-200 billion gallons each day." – Water: A Resource in Crisis by Eileen Lucas.

As this isn't a statistic for individuals, let's not include it in our tally.

32. <u>Thermoelectric Power</u> – (Page 231, IRWD) – There is a significant loss of water through evaporation and pollution. The vast majority of the water goes further downstream and is re-used. As this isn't a statistic for individuals, let's not include it in our tally.

33. <u>Other Miscellaneous Water Uses</u> – Water treatment systems, humidifiers, swamp coolers, leg shaving. As statistics on this are difficult to find, and this type of water usage is not the norm, let's not include it in our tally.

C. My Personal Consumption for the Previously discussed 33 Categories

(Please note, my water circulation system is not pressurized and is low flow.)

Sorry to bore you with this reminder but it's well documented. We tend to *grossly* underestimate the amount of water we use. As previously noted, according to a March 3, 2014 article in the L.A. Times, Americans use twice as much water as we thought.

I'm now presenting the same 33 water use categories, in the same order, as were listed in the last chapter. When applicable, their listings in the previous section have their reference(s).[2] Also feel free to peruse the book's bibliography. The hyperlinks to the biography are online at http://www.itvinstitute.com/wc-links.html

I'll now note the daily amount of water that I personally use for each of those 33 water use categories and compare it to the amount that the average American uses for each, which is noted in the previous chapter.

I sometimes use recycled water of my own making, and because that water would ordinarily be dumped out (but instead I re-use it,) I don't include it in my total tally of personal water use. Again that's because it's not water I am taking from the water table but instead water I have used once already and am re-using.

I live off-grid with the large majority of my electricity created by solar panels. The sun also heats much of my hot water (more on this further on,) though I also have a propane hot water system.

I use a solar oven to cook much of my food, and coffee, though a propane oven and stove is available. (Pictures of these are provided further on.)

1. Toilet Flushing – My average daily personal use is 1 gallon. As noted previously, <u>Americans use an estimated average of 18 gallons a day per person</u> for flushing the toilet(s). *But as I use recycled water for all my flushing I'm not including any of this in my final daily tally of water use.*

This 'recycled' water was originally in a drinking water pan for the animals that I put water out for. (More information on this in the "***Water for Pets and Other Animals to Drink***" section further on.) I clean the water pan out daily for the animal's sake then clean and use that water for flushing the toilet into the septic tank. Thus the animal's drinking water I would be disposing of ordinarily but instead I re-use it to flush with.

I use an RV toilet for going number two. It empties into a septic tank. This smaller toilet requires dramatically less water to flush. However I usually don't use it for urinating. Instead for urinating I go *downwind* in an outside area set up for urinating. (Incidentally it's amazing how much the honey bees like urine!)

Each flush in this RV toilet takes 1/3 – 1/2 gallon.

Urinating outside (and inconspicuously) doesn't need flushing.

Another option is having an outhouse.

2. Hand Washing – *My average daily personal use is **1.5 gallons**. As previously noted <u>Americans use an estimated average of 5 gallons a day per person</u> for washing their hands. My water system is not pressurized and is low flow.

As a single guy, I likely wash my hands less than average anyway but I do wash them many times a day.

3. Face Washing (separate from bathing) – *My average daily personal use is 1/3 gallon also.* As previously noted it's estimated that <u>Americans use an average of 1/3 gallon a day per person</u> for washing their face when not bathing.

My water system is not pressurized and is low flow. If I'm not taking a shower, I do wash my face on average a couple of extra times a week.

4. Hair Washing – *My average daily personal use is **0 gallons** of water.* (See "*Shower/bathing*".)

I wash my hair while showering. There is no additional water use for me than the 4.5 gallons a shower discussed further on in "*Showering/bathing*". I don't wash my hair separately.

5. Hair Washing Using Bottled Water – *My average daily personal use is **0 gallons**.* (I don't do this.) As this type of water usage is not the daily norm, let's not include it in our tally.

6. Shaving – *My average daily personal use is 0* **gallons** *a day*. As previously noted <u>Americans use an estimated average of 1 gallon a day per person</u> for shaving.

I use an electric shaver hooked up to my solar charging system, or I shave while showering. There is no additional water use for me than part of my 4.5 gallon shower discussed further on in *"Showering/bathing"*.

7. Leaving the Water Running When Cleaning Vegetables, fruit, etc. – *My average daily personal use is* **1/8 gallon**. (Yes I should eat more fresh vegetables.) As previously noted <u>Americans use an estimated average of 1/2 gallon a day per person</u> for this.

Remember you don't know if the picker/packager washed his/her hands before touching what's about to go into your mouth.

8. Teeth Brushing – *My average daily personal use is* **1/8 gallon**. As previously noted <u>Americans use an estimated average of 1 gallon a day per person</u> for brushing their teeth.

I use bottled water to do my rinsing and cleaning of tooth brush. No water is running while I wash my teeth.

9. Showering/bathing – *My average daily use is* **3.25 gallons**. As previously noted <u>Americans use an estimated average of 18 gallons a day per person</u> for showering/bathing.

I use a $38 Zodi brand battery powered portable shower which pulls water out of a mostly full 5 gallon food-grade plastic bucket. (I put 4.5 gallons of water in there.) Once that five gallon bucket is out of water, I don't have any more water to shower with. (While soaping and shaving I turn that portable shower off to save water.) I usually use the sun to heat my water. Five large glass jars painted flat black are filled with water, place on rigid foam insulation and covered with a double layer of 12 millimeter clear vinyl sheet over a PVC, metal or wood frame. (See photo 1.)

Photo 1. *My big glass former pickle jars that are painted flat black to heat my shower water. It gets steaming hot from at least a half day of good sunlight.*

I don't shower every day, particularly in the cooler months. I estimate I shower a total of 265 times a year.

4.5 gallons x 265 showers a year means I use 1192.5 gallons annually. Divide 1192.5 gallons of water by 365 and we find I use an average of about 3.25 gallons of water a day showering.

Any leftover shower water I recycle by putting into the animal's pan of drinking water. See *24, "Water for Pets and Other Animals to Drink".*

10. Drinking Water and Water Used in Cooking – *My average daily consumption of water/liquid is a 3/4 gallon a day.* (That does not include coffee. See #12.) As previously noted <u>Americans use an estimated average of 1 gallon a day per person</u> for drinking and cooking.

I buy my drinking water.

TIP – I re-use my cooking water when I can. For instance when there's leftover water from cooking noodles, I don't just dump it but often save it and use that water to cook something else (assuming the water hasn't spoiled.)

11. Laundry – *My average daily use is **1.5 gallons**.* As previously noted <u>each Americans uses an estimated average of 5.5 gallons a day per person</u> for laundry washing.

I'm single and live, and often work, alone. Neither of my jobs generally require fancy clothing, so a nice wardrobe is typically not necessary. I don't change my clothes too often and frankly don't wash my clothes as much as most. I wear one outfit for office and general living, and typically another outfit for physical work. I keep wearing each until they smell and/or are too dirty.

I use a Laundromat for serious washing. I go there about once every 1½ to 2 months. While there I use four washers, one or two of them are the extra large ones. I only use the recommended water conserving front loading washers.

TIP – Only doing partial loads wastes water!

However I hand-wash my work clothes in-between (and hang them on a clothesline to dry.) Using tap water I hand-wash several articles of clothing at the same time, sometimes after they've been soaked in a bleach solution overnight.

After a shower I usually put the towel and floor mat out on the clothesline in the sun for drying and sanitation. The sun's rays can be a sanitizer and I also bake/dry various pots (after they're cleaned) in sunlight for a long time to help with their sanitation.

12. Coffee Making and Coffee Pot Washing – *My average daily use is **1/2 gallon** of water.* As previously noted <u>Americans use an estimated average of 1/4 gallon a day per person</u> for coffee making and coffee pot washing.

I cook my coffee by putting a gallon of drinking water in a thin metal covered pot which has been painted flat black. (See photo 2.) It's all enclosed with a double layer of 20 millimeter thick clear vinyl sheet over a skeleton of wood dowels or ½ inch PVC pipe. To sterilize the pot I first clean and dry it then let it sit empty in that same enclosure for at least a day. As already noted I primarily use a solar oven for more efficient cooking.

I have two pots and two solar cooking set ups like in the photo, so coffee is usually available.

Photo 2. *This is my solar coffee cooker. Most of the day in the sun like this and 1-2 gallons of coffee cook up fine. Other things can be cooked in it this way if whatever is being cooked is thin. Liquids are better but cubed potatoes didn't cook well. Cubed potatoes cook fine in my solar oven (with enough sun.) See Photo 3.*

13. Dishwashing – *My average daily use is* **1/4 gallon** *of water.* As previously noted <u>Americans use an estimated average of 4.5 gallons a day per person</u> for dishwashing by hand (which is the type of dishwashing I do.)

For better sanitation I normally eat using disposable bowls, plates and eating utensils, though sometimes a sharp knife or firmer cooking utensil is utilized. 75%+ of my cooking (less in the winter) I do in my three year old solar oven. (See photo 3 further on.) As previously noted, solar ovens spread the heat out evenly, thus it's difficult to burn something when cooking in a solar oven, unlike a conventional oven. (Because of that the pots used in a solar oven typically need less water to be washed.) Also, because of this more efficient cooking, cooking things in their containers (containers such as plastic bags and metal cans) is easier and is something I do quite a bit. This of course means there are less pots and pans to wash as the food is not taken out of its packaging near as much as when a conventional oven or burner is used.

Photo 3. *My solar oven is three years old and still can reach 300 degrees F. if the conditions are right. It cooks most foods if there is enough sun. It also comes with a sun reflector to make it hotter (which isn't pictured.) It works fine for most of the year, assuming there is at least some strong sunlight. Rice and noodles often don't need boiling water to cook in it as the heat is so evenly spread out. It's a superior way to cook (though not as convenient.) It is very tough to burn food in a solar oven of this type as the heat is so evenly spread, unlike a normal oven where the heat comes from one or two sides.*

14. Utility Sink – *My average daily use is **0 gallons** of water.* As this type of water usage is not the daily norm, let's not include it in our tally.

15. Thawing – *My average daily use is **0 gallons** of water.* As this type of water usage is not the daily norm, let's not include it in our tally.

16. Garbage Disposal and Cleaning Out the Kitchen Sink – *My average daily use is **0 gallons** of water.* As previously noted <u>Americans use an estimated average of 1/2 gallon a day per person</u> for garbage disposal operation and cleaning out of the sink.

I don't have a garbage disposal and generally clean out the kitchen sink by using recycled (unused) shower water.

17. Cleaning/clearing Out Other Household Drains – As this type of water usage is not the daily norm, let's not include it in our tally.

18. Car Wash – *My average daily use is **1/4 gallon** of water.* As previously noted <u>Americans use an estimated average of 1/2 gallon a day per person</u> for car washing. I rarely wash my car. I typically let the rain wash my car. (Admittedly rain washing doesn't always work well if there is a lot of dust in the air.) Sometimes I do take it to a coin operated self-service car wash facility.

19. Hosing off the Garage, Driveway, Roof, Floors, House, Solar Panels and/or Sidewalk – *My average daily use is **1/4 gallon** of water.* As previously noted <u>Americans use an estimated average of 1.25 gallons a day per person</u> for hosing off the garage, driveway, roof, floors, house and/or sidewalk. I do hose off/wash my solar panels periodically, particularly in the drier months. See Photo 4.

Photo 4. *Those are four of my Unisolar 144 and 136 watt solar panels, on a support I built. These tough plastic Unisolar solar panels look like solar mats. I've found them to be real durable and productive. Sandbags hold the big white tarp down.*

Building Tip - The tarp has air pushing up on it from the inside of the structure and on the inside I have had to cover the wood skeleton that the tarp lays on, with light weight ¾ inch hard foam insulation, the type that line walls to the outside in homes. This blocks the air that's pushing up on the tarp from the inside.

20. *Hosing Myself, Kids or Others Off With the Garden Hose* – My average daily use is **0 gallons** of water. As this type of water usage is not the daily norm, let's not include it in our tally.

21. *Sprinklers and Home Irrigation Systems* – *My average daily use is 0 gallons of water.* As previously noted <u>Americans use an estimated average of .6 gallons a day per person</u> for this. I don't have a lawn or garden.

22. *Leaks* – *My average daily loss is 1/8 gallon a day.* As previously noted American homes lose an average of 31 gallons. <u>(Assuming a home has 4 people, that's a lose of 7.75 gallons a day)</u> from leaks. I have dramatically less opportunity to 'leak' water due to having less piping and my overall conservation.

23. *Running of Tap Water to Cool it Off for Drinking* – *My average daily use is 0 gallons of water.* I keep water in the cooler/refrigerator instead. As this type of water usage is not the daily norm, let's not include it in our tally.

24. *Water for Pets and Other Animals to Drink* – *My average daily use is 2 gallons of water.* (But as ½ gallon of it is recycled [unused] shower water, *I'm only putting 1½ gallon in my actual tally.* This is because I would have dumped out the unused ½ gallon of unused shower water if I didn't use it in this manner.)

I don't have any pets, or raise animals for consumption, but I provide water for the numerous wild animals and birds in the area. Every day I clean out their water pan and refill it. This takes about two gallons of water. However as I often have leftover shower water for recycling, I first use that. As many different animals drink from this water pan daily, to be on the safe side, I clean and re-fill the pan out daily, and that uses a significant amount of water. See photo 5.

In the animal's water pan I might also put rocks to take up volume and save water.

As noted earlier, I save, strain the leaves and bugs out of, and re-use (thus recycle) the un-drunk water from the animal's water pan for flushing the toilet into the septic tank.

Photo 5. *Wild turkeys and turkey chicks drinking from the wild animal drinking pan.(More chicks up against the boat.)*

25. Extra Water Usage Associated with hosting Parties and Other Social Events – **Undetermined.** I know this is taking the easy way out but I couldn't find data supporting a good enough estimate. It likely is a significant amount though as we use water preparing for guests.

26. Putting Water in the Car Radiator – As this type of water usage is not the daily norm, let's not include it in our tally.

27. Public Supply and Facilities – (*Such as Community Pools, Water Fountains, Restaurant Toilets and Patron's Complimentary Drinking Water*) – As a citizen of a community I use community and business facilities, which mean direct and indirect water use. For instance I eat sometimes at dining facilities such as fast food outlets and restaurants. As you know these provide a drink or glass of water and use liquid to make the meals we order. I don't use community pools, but do sometimes use the rest rooms of businesses and parks. Let's not include this in our tally.

28. Agricultural Use (*including irrigation and sprinklers*) – As previously noted <u>Americans use an estimated average of 5 gallons a day per person</u> for washing their hands. I do not grow plants for consumption (or anything else.) I don't have a lawn or potted plants. Let's not include this in our tally.

29. Industrial Aquaculture – I'm not involved with that. Let's not include this in our tally.

30. Mining – I'm not involved with that. Let's not include this in our tally.

31. General US Industry & Manufacturing – I'm not involved with that. Let's not include this in our tally.

32. Thermoelectric Power – I'm not involved with that. Let's not include this in our tally.

33. Other Miscellaneous Water Uses – **My average daily use is 1/2 gallon of water.** As previously noted Americans use an estimated average of 5 gallons a day per person for washing their hands. Let's be on the safe side though I think that amount is generous.

[1] That total is thought to have gone down due to conservation measures.
[2] In this report there are some estimations based on referenced data.

4. Conclusion

While I use dramatically less water than most Americans the obvious trade-off can be luxury, ease of operation, social status and needing more time to perform some tasks. For instance there is the extra effort of taking the shower water I don't use, carrying it to the animal's water pan and pouring it in. Also the animal's drinking water gets dirty from leaves falling in it so it needs to be cleaned out before recycling it as water for flushing the toilet.

There is no doubt that this low carbon, extreme water conservation life style is going to be tough for most, particular those with children. A lot of that is because people associate this life style with poverty. My personal experience is that this is a particularly hard sell for women, particularly those with children and/or past her 20s.

Another concern would be sanitation. To my knowledge I have not directly gotten ill from this extreme water conservation life style. I have however had issues with living in the wilderness, such as being attacked by a feral (stray) cat.

Also there is the question of what do I get out of it? For one thing I save money. Hopefully now I also have written a quality book for you to read. In real life though my impression is that while most think it's "cool", "interesting" and "good that someone's doing it for everybody's sake," they themselves typically don't want to do it for any serious length of time. This is particularly true of women. Again, I think a lot of it is that society thinks people who live like this are of a lower caste (so to speak), even though I don't live in squalor.

For those interested, I live alone in North Central California. I'm not a hippie, never been arrested, not wanted by the law, am healthy (for my age) and have a four year college degree. Thank you.

5. Bibliography and Other Reference Material

The following hyperlinks are listed online for easier access.
http://www.itvinstitute.com/wc-links.html

"The US Geological Survey." Water use by type in 2005, as reported by Kenny, J.F., Barber, N.L., Hutson, S.S., Linsey, K.S., Lovelace, J.K., and Maupin, M.A., 2009, "Estimated Use of Water in the United States in 2005".

http://www.epa.gov/WaterSense/our_water/start_saving.html
http://www.epa.gov/WaterSense/pubs/indoor.html
https://www.garrettcounty.org/public-utilities/general-information/water-conservation-tips
http://www.allianceforwaterefficiency.org
http://www.reuters.com/article/2015/05/23/us-usa-california-dam-idUSKBN0O804M20150523
http://www.irwd.com/images/pdf/save-water/CaSingleFamilyWaterUseEfficiencyStudyJune2011.pdf
http://water.usgs.gov/edu/qa-home-percapita.html - a lot of statistics.
http://sustainability.about.com/od/Sustainability/a/How-We-Use-Water.htm
http://ww2.kqed.org/lowdown/2014/01/23/how-much-water-do-californians-use-each-day-and-what-does-a-20-reduction-look-like/
http://www.epa.gov/watersense/our_water/water_use_today.html
http://listserver.energy.ca.gov/appliances/2013rulemaking/documents/responses/Water_Appliances_12-AAER-2C/California_IOU_Response_to_CEC_Invitation_to_Participate-Toilets_and_Urinals_REFERENCES/Aquacraft_1999_Residential_End_Uses_of_Water_Study.pdf
http://www.stormwater.cecs.ucf.edu/toolkit/vol1/Contents/pdfs/Water%20Conservation/Water%20used%20handout.pdf
http://www.mercurynews.com/data/ci_25059942?appSession=681116276930838&RecordID=&PageID=2&PrevPageID=2&cpipage=1&CPIsortType=asc&CPIorderby=Pop2010&cbCurrentPageSize=250&appSession=478116276787156
http://www.epa.gov/WaterSense/our_water/learn_more.html#tabs-6 – a lot of links to statistics
http://www.swfwmd.state.fl.us/publications/files/daily_water_use.pdf

The End

Book #3
The Ultimate Collection of Resurrections and Rebirths

This publication is designed to provide accurate and authoritative information in regard to the subject matter covered. It is sold and/or distributed with the understanding that the publisher and author is not engaged in rendering legal or other professional services. In this publication there may be inadvertent inaccuracies including technical inaccuracies, typographical inaccuracies and other possible inaccuracies. **The writer and publisher of this publication expressly disclaim all liability for the use or interpretation by others of information contained in this publication** and/or listed Web sites. **The author, publisher and distributors of this publication hereby disclaim any and all liability for any loss or damage caused by errors or omissions, (should these exist) whether such errors or omissions resulted from negligence, accident, or other causes.** If legal advice or other expert assistance is required, the services of a competent professional person in a consultation capacity should be sought. This publication may list Web links to Web sites/pages which are not maintained by any party or parties involved with this publication. Products, services and websites' content vary with time. Please verify any published information.

Hardcopy ISBN-13: 978-1489584588
Hardcopy ISBN-10: 1489584587

The Ultimate Collection of Resurrections and Rebirths

Table of Contents

1. List of Famous Resurrections and Rebirths

The following is a list of historically famous resurrected people and deities from tradition, mythology and other text. There may be more than one version of events for some of the below.

Please note that not everybody agrees as to their resurrection/rebirth and presented specifics. Further research on each of the following is advocated before making a definitive statement as to its regard.

(In regard to an entry with a question mark after them, I was not able to get adequate confirmation information. You however might be able to.)

ADONIS - (Greece) **(1)** Venus saw him hunting and fell in love with him. This offended Mars, the rival for the affections of Venus, and he assumed the form of a wild boar and killed Adonis who was hunting. Venus mourned exceedingly. She was so overcome with grief and fear that she went down into the lower regions (underworld) to bring back her lover (Adonis). But Pluto's wife saw how handsome Adonis was and she would not let him go back up to earth with Venus so they came to an agreement where they would divide the year into halves, and each in turn should have him for a half. **(2)** A very handsome boy who Zeus was persuaded by the goddess Aphrodite to resurrect each year for 6 months.

AENEAS – (Rome) – According to the ancient writer Virgil, Aeneas is devastated by Troy being ruined. Aeneas dies (goes to the underworld) and comes back as a new, determined more optimistic Aeneas.

ATUNIS – See the Greek God *Adonis*.

ALCESTIS - When it came time for the Greek deity Admetus to die, Apollo arraigned for someone else to die for him. The deity Alcestis volunteers and died for him, but is resurrected after other gods decided that such an arrangement was unethical.

ASSUR - Assur is thought to be the pre-Egyptian king and conqueror of a huge amount of the ancient world, including what would become Egypt.

UrRea was an agrarian civilization in the Nile Delta before there was an Egypt. Storytellers handed down the history of that period for countless generations before there was writing.

We're told that Assur cleared jungles, drained swamps and dug canals. He determined which plants were fit for food and medicine. He taught his people agriculture and animal husbandry. He started the first kingdom, made laws, taught people to worship the gods and even ended cannibalism. After he had raised his own people up from savagery, he went about civilizing the whole world.

Assur's wife was the legendary Isis. One day the evil god Set, and other conspirators killed Assur. They cut his body into pieces and scattered his severed parts in many places.

Assur's faithful wife Isis gathered his scattered body parts and put them all back together. She also resurrected Assur from the dead. After his resurrection, they had a son, Heru. Assur would become the king of the afterlife and judge of the dead. Isis went on to become the EarthMother Goddess. (Isis thus could be considered the original Eve.) Heru went on to become the progenitor of all the Pharaohs and the patriarch of Egyptian civilization. Set went on to become Satan, the god of evil of the Egyptian religion. All the Pharaohs traced their lineage to the gods, Assur, Isis, and Heru.

Note: The people of UrRea were UrReans, not Africans. They spoke UrRean languages, not Egyptian. They didn't become Egyptians until after the birth of Isis and the resurrection of Assur. This pre-Egyptian UrRean civilization in the Nile Delta was related to Pre-Sumerian and Pre-Akkadian civilizations between the Tigris and Euphrates rivers.

Assur is known as "Osiris" in conventional mythology. The ancient Greeks translated his name to be that. When Alexander the Great conquered Egypt, that land was called "Mis-Ur-Re" thus Egypt was called *Mis-Ur-Re* before it was called Egypt. The Greeks changed the name of the country to "Egyptos". Then they gave the Egyptian gods Greek names. Osiris was Assur for thousands of years before he was called Osiris. Heru's name would be translated by the Greeks into "Horus".

ATTIS - A Mesopotamian (Phrygian) god. He sacrificed himself to Zeus (called Jupiter by the Roman Priests,) but is resurrected every spring.

BAAL – Baal (or Ba'al) means "master" or "lord". Many deities in various parts of the world were called Baal (or Ba'al). Ba'al of Tyre is Melqart. See *Melqart*. In this version of events, Baal goes down to the underworld and dies. In a vision the chief god, El, sees that Baal has returned to life and Baal returns to his throne.

BACCHUS - The Roman god of wine and intoxication. It's Greek equivalent is Dionysus. See *Dionysus*.

BALDR (also Balder, Baldur) – A famous figure in old Norse/Danish poetry. One account has Baldr mistakenly dying from the arrow, (another account has it as a spear) of his blind brother Barbarika. His death led to the destruction of the gods at Ragnarök and according to Völuspá, Baldr will be reborn in the new world.

BARI – (Korea) - This is a myth regarding the origin of the shamans. The ancestral shaman is believed to be Bari, a sacred woman. She works many miracles. She goes down into the underworld (where the dead go) and not only frees people from it but returns to life up here. Her royal parents die and she brings them back to life. Bari became the death goddess, the guider of the dead to the Underworld. She also became the first shaman, and the patron of all the shamans in Korea.

BUDDIAH (Siddhartha Gautama) - He along with Horus, Mithra, Dionysus and Osiris have the most glaring similarities to Jesus' life as we know it. Buddha was born to the Virgin Maya on December 25th. (Not all agree on this date of birth.) His birth was announced by a star in the sky and his birth drew wise men presenting costly gifts. He was baptized in water. Buddha healed the sick and fed 500 from a small basket of cakes and even walked on water. He died (on a cross, in some traditions,) was buried but arose again after his tomb was opened by supernatural powers. He ascended into heaven but is expected to return in later days to judge the dead and certain others. As noted, Buddiah was resurrected when his tomb was opened by supernatural powers.

CASTOR & POLLUX - Sons of Zeus though some accounts have them as mortals. When Castor was killed, Zeus was quite unhappy and decided to have each son live on alternating days. Pollux is given two options by Zeus and this is the one he chose.

CHINNAMASTA - (Northern India and Nepal) - Translated to be "She whose head is severed". Chhinnamasta is recognized by both Buddhists and Hindus. She's equated with Chinnamunda – the severed-headed form of the Tibetan Buddhist goddess Vajrayogini. Her head was cut off and she came back to life (or never died) and lived on. Among other things she is the goddess of sexual desire.

DIONYSOS - (Greece) - This is a Phrygian and later adopted Greek god. He along with Horus, Mithra, Buddha, Krishna and Osiris have the most glaring similarities to Jesus' life as we know it. He was born of a Virgin on December 25th or January 6th. (December 25th would be the date the ancient

Greeks would settle on.) He was placed in a manger. (Mangers were a lot more of a commonplace in those days since there were so many horses. Mangers were where many poorer travelers were allowed to stay.) He was a traveling teacher who performed many miracles. He turned water into wine. His followers could eat a sacred meal. He is said to have rose from the dead on March 25th. He was called the "Redeemer," "Sin bearer," "King of Kings," "the Anointed One," "Only Begotten Son," "Savior."

DUMUZI - Tammuz was established in honor of the eponymous god Tammuz, who originated as a Sumerian shepherd-god, Dumuzid or Dumuzi. See *Tammuz*.

ESHMUN - (or Eshmoun, less accurately Esmun or Esmoun) – Eshmun was a Phoenician god of healing and the guardian or protector god of Sidon. The goddess Astronoë so harassed him with her love for him that in desperation he castrated himself and died. Astronoë resurrected him from the warmth of her body, and changed him into a god.

EURIDICE - (Greek) When Orpheus' wife Euridice dies of a snake bite, the grieving Orpheus went down to Hades (where all dead Greeks went according to tradition) and persuaded the god in charge, Pluto, to resurrect Euridice. Pluto promises to resurrect Eurydice if Orpheus doesn't look back at her as Orpheus heads back to earth, but he does and she doesn't make it past the underworld.

FRODE - Frode is the name of a number of legendary Danish kings including Beowulf. One version has a Frode disappearing into the earth for three years after he dies and then returning.

GULLVEIG - In Norse mythology, Gullveig is a speared, then burned three times. He however keeps getting reborn after each burning.

HAY-TAU - Egyptian vegetation god. He died and miraculously came back to life (with no other god's help) several years later.

HEITSI-EIBIB - (Saan, or Bushmen of Southern Africa) - Heitsi-eibib is said to have died and resurrected himself on several occasions. Because from this his funeral cairns (man-made pile [or stack] of stones) are found in many locations in southern Africa and it is customary to throw (or put) a stone onto them for good luck.

INANNA - The Sumerian goddess of sexual love, fertility, and warfare. She is the equivalent of the better known *Ishtar*. Inanna is reborn after she travels to the underworld and dies there. She descended into the land of the dead (underworld) which was ruled by her sister, Ereskeigal, the goddess of death and infertility. She went down there to rescue her lover, a vegetation god named Tammuz, who was being held hostage. In this version of events she is brought back to life when her servant sprinkled her with the "water of life."

IRAVAN – (South India) - (aka Aravan, Iravat and Iravant) - He is a character from the Hindu epic Mahabharata. One version is that the great Krishna allows Iravan to witness the entire duration of the Mahabharata war through the eyes of his severed head.

ISHTAR - East Semitic Akkadian, Assyrian and Babylonian goddess. In earlier Sumerian culture she was known as Inanna. See *Inanna*.

IZANAMI - (Izanami-no-kami and/or Izanagi no Mikoto) – (Japan) - ?

JESUS CHRIST – Jesus was resurrected.

JARILO – (Yarilo, Iarilo, or Gerovit) – (Slavic) – Jarilo is believed to be (re)born and killed every year. His mythical life cycle followed the yearly life of various types of vegetation.

KOSTROMA - (Slavic) – ?

KRISHNA - (Chrishna) - The mother of Chrishna (Devaki) was "overshadowed" (taken sexually) by the supreme god, Brahma. Chrishna's mother had given birth seven times before but still remained a virgin (as was the case with Mother Mary who gave birth to 3 boys, including Jesus, and also still remained a virgin.) Krishna's birth was announced by a star. Krishna was born in a cave, which at the time of his birth was miraculously illuminated. King Kansa sought the life of the Indian Christ by ordering the massacre of all male children born during the same night as Krishna. Krishna traveled widely, performing miracles, including raising the dead, healing lepers, healing the blind and the deaf. The crucified Krishna is pictured on the cross with arms extended. Pierced by an arrow while hanging on the cross, Krishna died, but descended into Hell from which He rose again on the third day and ascended into Heaven. (The Gospel of Nicodemus tells of Jesus' descent into Hell.) He is expected to return on the last day. Chrishna's birthday is often thought to be December 25th. (Others claim that Krishna's birth was in July, August or September.)

LEMMINKÄINEN - Lemminkäinen is a prominent figure in Finnish mythology. He is one of the Heroes of the Kalevala. Lemminkäinen dies and in an effort to reassemble him, his mother searches all over for all parts of his body. Finally, with help, she is able to get him to live again.

MARZANNA – (Mara, Maržena, Morana, Moréna, Mora or Marmora) - Slavic goddess associated with death and rebirth of nature including seasonal agricultural growth and death.

MELQART – (Melqart-Heracles, Herakles) – Melqart's Greek equivalent is Heracles) - This Phoenician god is resurrected. Melqart is respected and/or worshiped in Phoenician and Punic cultures from Syria to Spain. He is the god of the Phoenician city of Tyre. Heracles was killed by a Typhon and raised from the dead by Iolaos.

MITHRA - "Christianity's sister religion." He along with Horus, Osiris, Krishna, Dionysus and Buddha have the most glaring similarities to Jesus' life as we know it. Mithra had a virgin birth of a sort, Mithra was said to be born from the "rock of a cave" while shepherds and many Magi looked on.

Mithra came to be known as the all important sun God. He originally was a Zoroastrian God who people began to associate independently of Zoroastrianism. *Mithraism is often called Christianity's "sister religion" largely because there are so many similarities between the two and it originated and was popular prior to Christianity.* Mithraism spread into Greek and Roman influenced areas from Persia, coming to Rome in 68 B.C. It, and the worship of Greek and Roman Gods, were Christianity's biggest competition until the latter 3rd century. In fact, until the latter 3rd century, it was a more popular religion than Christianity.

Many of Mithraism's doctrines are strikingly similar to Christianity. In particular, communion, the use of holy water, the adoration of shepherds at Mithras' birth, the use of Sundays for the Sabbath instead of the traditional Saturday which is the Sabbath of the Jewish people, and using the date of December 25th as the birth date of both Jesus and Mithra, (it had been celebrated as Mithra's birth date for many years before Christians adopted it for Jesus.) Also the Mithraic belief in the immortality of the soul, the last judgment and the resurrection preceded Christianity's adopting those beliefs. However, it's different from Christianity in significant ways also. Mithra was considered a God not a prophet. Its secretive ceremonies excluded women. They also had other lesser gods.

Mithraism was quite popular with Roman soldiers. Though it preached brotherly love, as with Islam and Sikhism, it wished to aid the soldier in battle. People becoming Mithraistic were often baptized in the blood of a bull. After Christianity became Rome's official religion, Christians destroyed most of the Mithratic temples and killed any Mithratic clergy they could find.

Mithraism exposed millions of people to doctrines and procedures which would later be preached by the newcomer, Christianity. Its similarities with Christianity may have made it easier for Christian doctrines to have been so quickly and readily accepted. Had Christianity not been so successful, it's possible Mithraism would have become the official religion of Rome.

Mithra was resurrected after death but he comes back as another person.

ODIN – (Norse Mythology) – Odin is a major Norse god and was popular with the Vikings. Odin lost an eye. His love for wisdom however was so strong that he was willing to sacrifice his life to try and find more knowledge of the world. He gained insight by hanging himself for nine days from Yggdrasil, the cosmic tree. Following his voluntary death he had a magical resurrection.

OBATALA - (Yoruba people) - Obatala is a dying and rising god. He left his Temple on the seventh day of the Itapa festival, stayed in his grave on the eighth day and returned in a great procession to his Temple on the ninth day.

ORION – Associated with Osiris. See *Osiris*.

ORPHEUS - Orpheus was a legendary poet, prophet and musician in ancient Greek religion and myth. In his quest to get his wife Eurydice reborn after her untimely death, he was one of the handful of Greek heroes to visit the Underworld and returned alive (thus he didn't so much die but went to the underworld and returned alive.)

OSIRIS - The Egyptian god who was murdered and resurrected. He however was resurrected only in the after world, thus becoming the very important Egyptian god (and king) of the afterlife. See "*Assur*". "Osiris" was the translated name given to Assur" by the Greeks who in many ways adopted Assur.

OUROBOROS - An ancient symbol showing a serpent or dragon eating its own tail. It is a symbol of resurrection.

PERSEPHONE - Persephone was abducted by Hades the god-king of the Greek underworld. She would become his wife making her the queen of the underworld. Because Persephone had tasted of the food of Hades she was forced to forever spend a part of the year with her husband in the underworld.

Her annual return to the earth was in the spring and that made her the Goddess of spring. Her return to the underworld in winter, conversely was associated with the dying of plants until she returned to the upper world in spring.

PHOENIX - A phoenix obtains new life by arising from the ashes of its predecessor.

PROSERPINA – (Proserpina, Proserpine, Persephone) – She is an ancient Roman goddess whose story is the basis of a myth of springtime. Her Greek goddess equivalent is Persephone. Pluto, Roman king of the underworld, came out of the underworld and abducted her to live with him down in the underworld. She would eventually be freed but would have to live three months of each year with Pluto, and stay the rest with her mother who wasn't in the underworld.

QUETZALCOATL - Quetzalcoatl was born of a virgin around 900 BC. His story is told by the Aztecs and Mayans. He was crucified (or sets himself on fire after over-drinking depending on the version.) He was considered the God of light that struggled against the god of darkness called Tezcatlipoca. He was associated with the planet Venus, the morning star, like Jesus is. The cross was used as a symbolic representation. He said he would return from the dead to claim his earthly kingdom.

RA - Ra was identified primarily with the sun and is a major god in the religion of ancient Egypt. Humans were created from Ra's tears and sweat. Ra would go to the underworld at night (which is largely why it got dark) and be is reborn daily for the sun to rise.

TAMMUZ - (Arabic) - An ancient Sumerian-Akkadian god who dies and is resurrected every year. Many religious historians consider him the great grand daddy of all religious resurrections. The Sumerians are thought to have worshipped him (and his resurrection) as far back as 4,000 BC. (2) His Greek equivalent is "Adonis". See *Adonis*.

XIPE TOTEC – The Aztec god of rebirth and more. He doesn't get reborn as such, he flays his skin regularly and it grows back with no problem.

ZALMOXIS – Zalmoxis is another Life-Death-Rebirth deity. Some see him as a figure that resurrects.

ZOROASTRIANISM - While it currently has only an estimated 250,000 believers (1990 statistic), primarily in India and Iran, this religion and it's offshoot religion, Mithraism, has had a tremendous impact on the development of Christianity and Judaism.

Founded by the Persian prophet Zoroaster sometime between 1000 and 600 B.C., it's one of the world's original *dualistic* religions (see glossary for definition of dualism). It was practiced most by people in Arabia, and south-central Asia. Verifiable history shows it flourishing in the Persian Empire by 550 B.C.

After Alexander the Great's invasion of Persia (334 B.C.,) Zoroastrianism lost most of its followers. It wouldn't experience a serious revival until hundreds of years later. Zoroastrianism once again lost popularity when in roughly 637 A.D., Islamic forces conquered much of the territory it was practiced in, forcibly replacing it with Islam.

Like Christianity, Zoroastrianism is dualistic. It preaches that there is a struggle between good and evil forces for mastery of the universe. The good God and creator of the universe, is Ahura Mazda. The evil demon is Ahriman and he's bent on destroying everything good in the universe.

Zoroaster was sent down by God (Ahura Mazda) to help humans fight evil. He received the sacred laws of God, the Avesta on Mount Sabalan, in Persia. If an individual were to choose the path of goodness which Zoroaster represents, then on Judgment Day, he/she would go to heaven. At some point, a great battle between God, his angels (the "Amesha Spenta") and those "good" converted humans will occur against the evil demons and "bad" people who have not "converted." This will be the final battle and be the final "Judgment of the Dead*." Since this was preached, at least in Jesus' area, six to ten centuries before Jesus was even born, it's history's original "Judgment Day."* Peace will reign on earth after this battle because it is now cleansed of evil (unlike the Christian version of Judgment Day which assumes it is the end of the world.)

Many similarities exist between Zoroastrianism, Judaism and Christianity. Both Zoroaster and Moses got their respective religious "Laws" from their one true God on a mountain (he getting the Avesta from Ahura Mazda on Mt. Sabalan and Moses getting the Judaic "laws" from Yahweh on Mt. Sinai.) Each is dualistically fighting a power of evil, Ahura Mazda against Ahriman, and Yahweh and the God of Christianity against Satan. The Zoroastrian all important first period of creation is divided into 6 parts as the Old Testament account of creation is divided into six days. A human couple is the first human beings for each religion, Moshya and Moshyana by the Zoroastrian account and Adam and Eve by the other two religious accounts. A terrible winter was sent by Ahura Mazda, as told in the Avesta, to punish mankind for its evil. This, of course, parallels the great flood theory. All three religions believes in a good life after death for the righteous and a hellish life after death for the unrighteous. A Zoroastrian, during a religious ceremony, drinks milk and water and eats bread. Originally, Christians did the same with wine and unleavened bread (now it's more often wafer.) Zoroastrians celebrated high and low masses. A religion worshipping the lesser Zoroastrian God of light, Mithra, spread from Persia to the Mediterranean and even Rome. In fact, in Rome, it was a more popular religion than Christianity for almost the first three centuries A. D. Various Christian assertions including the date of December 25th for the traditional birthday of Jesus, comes from Mithraism (Mithra's birthday was celebrated on December 25th for many years prior to Jesus' birth.) For more information on what's called *Christianity's sister religion*, see the description of Mithraism in this section.

2. Afterlife from the Standpoint of Science

There are at least five types of scientifically proven afterlife scenarios.

1) First off, genetics. When you procreate, a large portion of your DNA survives you in the form of your children, direct relatives and to a certain extent, your fellow human beings. If you have the DNA which make men lose their hair, that DNA is passed on and will probably be noticeable in generations preceding you. This is true, of course, about many physical, emotional and mental characteristics, and even diseases such as diabetes.
People might be surprised how long an afterlife (at least physically) we have thanks to our DNA. Even though your son may not look or seem anything like your great, great, great, great, grandfather, present in your son's body is some of his DNA.

Scientists say we all come from a few original human beings. You may say this supports the theory of Creation but from a scientific standpoint it doesn't. First off, the earliest man-like creatures may have been asexual, being able to procreate all by themselves. (Sometimes referred to as hermaphrodites.) This simply was a bad idea from the standpoint of evolution for the same reason that it is not a good idea for most relatives to procreate. Everybody's chromosome chains have weak links. Family members often have the same weak links. These weak links need to be substituted with stronger links. These stronger links are best supplied by a totally different family tree. Asexuality, while convenient, possibly would have doomed our species to countless mutations and unstable life forms.

2) Another form of afterlife is one's posthumous impact on other people. This impact doesn't have to be mankind as a whole but could be on a much smaller "group of people" such as one's family. If the family is still talking about your great, great, great grandmother, even occasionally, she would qualify for this type of "afterlife." It can be said then that Jesus has had a phenomenal afterlife. Others with impressive, though less spectacular afterlives include Alexander the Great, Abraham Lincoln and Galileo.

3) Nuclear physics tells us that our body's matter (atoms and molecules,) also has an afterlife. As an example, a dead body buried in the ground eventually provides nourishment for countless organisms, mostly microscopic. Perhaps a part of one's body is used in an organ transplant. This lies in the realm of physics known as Matter Transference.

4) Still a fourth type deals with Cryonics. The freezing of the body until another time in the future when it can be thawed out and reborn.

5) There are those who feel if you stop breathing, and/or your heart stops beating, but you are brought back to life via mouth to mouth resuscitation, CPR or the like, you have been "resurrected" and technically are living in an afterlife.

3. Epilogue

There are rare documented cases of the return to life of the clinically dead. This is known as Lazarus syndrome, a term originating from the Biblical story of the Resurrection of Lazarus.

Many who have had Near Death Experiences (NDE) have, in their opinion at least, seen and/or in some other way experienced the afterlife. Seeing the light at the end of a tunnel and floating above their bodies has also been experienced by pilots who pass out from too powerful G-forces, particularly in the training centrifuges. Recorded data noted that their vital signs typically were still ongoing during the NDE 'experience'.

The End

www.ingramcontent.com/pod-product-compliance
Lightning Source LLC
Chambersburg PA
CBHW071002180526
45168CB00003B/1256